ASSESSMENT HANDBOOK

Everyday
Mathematics®

The University of Chicago School Mathematics Project

Mc
Graw
Hill
Education

The University of Chicago School Mathematics Project

Max Bell, Director, *Everyday Mathematics* First Edition; James McBride, Director, *Everyday Mathematics* Second Edition; Andy Isaacs, Director, *Everyday Mathematics* Third, CCSS, and Fourth Editions; Amy Dillard, Associate Director, *Everyday Mathematics* Third Edition; Rachel Malpass McCall, Associate Director, *Everyday Mathematics* CCSS and Fourth Editions; Mary Ellen Dairyko, Associate Director, *Everyday Mathematics* Fourth Edition

Authors
Max Bell, Jean Bell, John Bretzlauf, Amy Dillard, Robert Hartfield, Andy Isaacs, James McBride, Cheryl G. Moran, Kathleen Pitvorec, Peter Saecker

Fourth Edition Grade 2 Team Leader
Cheryl G. Moran

Writers
Camille Bourisaw, Mary Ellen Dairyko, Gina Garza-Kling, Rebecca Williams Maxcy, Kathryn M. Rich

Open Response Team
Catherine R. Kelso, Leader; Steve Hinds

Differentiation Team
Ava Belisle-Chatterjee, Leader; Jean Marie Capper, Martin Gartzman

Digital Development Team
Carla Agard-Strickland, Leader; John Benson, Gregory Berns-Leone, Juan Camilo Acevedo

Virtual Learning Community
Meg Schleppenbach Bates, Cheryl G. Moran, Margaret Sharkey

Technical Art
Diana Barrie, Senior Artist; Cherry Inthalangsy

UCSMP Editorial
Don Reneau, Senior Editor; Rachel Jacobs, Kristen Pasmore, Luke Whalen

Field Test Coordination
Denise A. Porter

Field Test Teachers
Kristin Collins, Debbie Crowley, Brooke Fordice, Callie Huggins, Luke Larmee, Jaclyn McNamee, Vibha Sanghvi, Brook Triplett

Digital Field Test Teachers
Colleen Girard, Michelle Kutanovski, Gina Cipriani, Retonyar Ringold, Catherine Rollings, Julia Schacht, Christine Molina-Rebecca, Monica Diaz de Leon, Tiffany Barnes, Andrea Bonanno-Lersch, Debra Fields, Kellie Johnson, Elyse D'Andrea, Katie Fielden, Jamie Henry, Jill Parisi, Lauren Wolkhamer, Kenecia Moore, Julie Spaite, Sue White, Damaris Miles, Kelly Fitzgerald

Contributors
William B. Baker, John Benson, Jeanne Mills DiDomenico, James Flanders, Lila K. S. Goldstein, Funda Gönülateş, Lorraine M. Males, John P. Smith III, Kathleen Clark, Patti Satz, Penny Williams

Center for Elementary Mathematics and Science Education Administration
Martin Gartzman, Executive Director; Jose J. Fragoso, Jr., Meri B. Forhan, Regina Littleton, Laurie K. Thrasher

External Reviewers

The *Everyday Mathematics* authors gratefully acknowledge the work of the many scholars and teachers who reviewed plans for this edition. All decisions regarding the content and pedagogy of *Everyday Mathematics* were made by the authors and do not necessarily reflect the views of those listed below.

Elizabeth Babcock, California Academy of Sciences; Arthur J. Baroody, University of Illinois at Urbana-Champaign and University of Denver; Dawn Berk, University of Delaware; Diane J. Briars, Pittsburgh, Pennsylvania; Kathryn B. Chval, University of Missouri–Columbia; Kathleen Cramer, University of Minnesota; Ethan Danahy, Tufts University; Tom de Boor, Grunwald Associates; Louis V. DiBello, University of Illinois at Chicago; Corey Drake, Michigan State University; David Foster, Silicon Valley Mathematics Initiative; Funda Gönülateş, Michigan State University; M. Kathleen Heid, Pennsylvania State University; Natalie Jakucyn, Glenbrook South High School, Glenview, IL; Richard G. Kron, University of Chicago; Richard Lehrer, Vanderbilt University; Susan C. Levine, University of Chicago; Lorraine M. Males, University of Nebraska-Lincoln; Dr. George Mehler, Temple University and Central Bucks School District, Pennsylvania; Kenny Huy Nguyen, North Carolina State University; Mark Oreglia, University of Chicago; Sandra Overcash, Virginia Beach City Public Schools, Virginia; Raedy M. Ping, University of Chicago; Kevin L. Polk, Aveniros LLC; Sarah R. Powell, University of Texas at Austin; Janine T. Remillard, University of Pennsylvania; John P. Smith III, Michigan State University; Mary Kay Stein, University of Pittsburg; Dale Truding, Arlington Heights District 25; Judith S. Zawojewski, Illinois Institute of Technology

Note

Too many people have contributed to earlier editions of *Everyday Mathematics* to be listed here. Title and copyright pages for earlier editions can be found at http://everydaymath.uchicago.edu/about/ucsmp-cemse/.

www.everydaymath.com

Send all inquiries to:
McGraw-Hill Education
8787 Orion Place
Columbus, OH 43240

ISBN: 978-0-07-902118-2
MHID: 0-07-902118-2

Printed in the United States of America.

3 4 5 6 7 8 9 LHS 21 20 19

Contents

Assessment in Everyday Mathematics®

Assessment in *Everyday Mathematics*:

- addresses the full range of content and processes/practices in the Standards for Mathematics
- consists of tasks that are worthwhile learning experiences.
- is manageable for teachers.
- informs instruction by providing actionable information about children's progress.
- provides information for grading.
- clarifies the *Everyday Mathematics* spiral and helps teachers decide when to intervene and when "watchful waiting" is appropriate.
- serves basic Tier 1 and Tier 2 Response to Intervention (RtI) functions.
- provides information that will complement data from standards-based assessments, including those from PARCC and SBAC.

Go Online for more information in the Assessment section of the *Implementation Guide*.

Assessment of Content and Process/Practice Standards

Everyday Mathematics integrates instruction and assessment of mathematical processes and practices with instruction and assessment of grade-level content. The mathematical processes and practices are not to be separated from the content; they are mathematical habits of mind children should develop as they learn mathematical content.

However, the content and process/practice standards differ in important respects. The content standards describe specific goals that are organized by mathematical strand and differ from grade to grade. The process and practice standards describe general, cross-grade goals that are related to processes and practices such as problem solving, reasoning, and modeling. Many tasks in *Everyday Mathematics* provide opportunities to assess both content and process/practice standards. However, due to the differing nature of these standards, *Everyday Mathematics* assesses and tracks progress on them in different ways.

Assessing the Content Standards

Each grade's content standards are unpacked into 45 to 80 *Everyday Mathematics* Goals for Mathematical Content (GMC). The standards and the corresponding GMCs are listed in the back of the *Teacher's Lesson Guide*. Instructional activities and assessment items are linked to one or more of the GMCs, which provide more targeted information for assessment and differentiation.

For each task that assesses a content standard, *Everyday Mathematics* provides guidance on what constitutes "meeting expectations" for that standard at that point in the year. Individual Profiles of Progress, Class Checklists, and the assessment and reporting tools help teachers monitor children's progress using this framework.

Assessing the Process and Practice Standards

Since the Standards for Mathematical Process and Practice are broadly written for Grades K to 12, *Everyday Mathematics* includes Goals for Mathematical Process and Practice (GMP) that unpack the standards for elementary school teachers and children. These *Everyday Mathematics* GMPs are useful for assessing the mathematical process and practice standards because they highlight specific aspects of each process and practice.

Tracking progress on the mathematical process and practice standards requires a more qualitative approach. Assessment opportunities include writing/reasoning prompts, open response problems, and observations of children in the course of daily work. Tools for assessing the processes and practices include checklists and task-specific rubrics for open response problems.

[Go Online] for more information about the GMCs and GMPs in the *Everyday Mathematics* and the Standards section in the *Implementation Guide*.

Assessment Opportunities

Everyday Mathematics offers a variety of opportunities for ongoing and periodic assessment of content and process/practice standards.

Assessment Check-Ins

Assessment Check-Ins are lesson-embedded opportunities to assess the focus content and processes/practices of the lesson. They appear in regular lessons and Open Response and Reengagement lessons. Each Assessment Check-In provides information on expectations for particular standards at that point in the curriculum. The results can be used to inform instruction and, often, for grading.

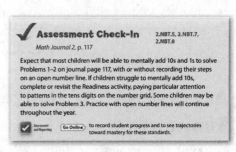

Preview Math Boxes

One pair of Preview Math Boxes appears near the end of each unit. The Preview Math Boxes can be used to gauge children's readiness for the upcoming unit so that teachers can better plan instruction and choose appropriate differentiation activities.

Writing/Reasoning Prompts

Many Math Boxes have Writing/Reasoning prompts that encourage children to communicate their understanding of concepts and skills and their strategies for solving problems. Writing/Reasoning prompts provide valuable opportunities for assessing the mathematical process and practice standards.

> **Writing/Reasoning** Explain how you made the largest possible number in Problem 3.

Progress Checks

The last lesson in each unit is a Progress Check, a two-day lesson that provides the following assessment opportunities:

- **Self Assessment** Every unit includes this opportunity for children to consider how well they are doing on the focus content of the unit.

- **Unit Assessment** Every unit includes this assessment of the content and process/practice standards that were the focus of the unit. All items are appropriate for grading because they match the expectations for the standards they assess up to that point in the year.

- **Open Response Assessment** Odd-numbered units include this opportunity for children to think creatively about a problem. It addresses one or more content standards and one Goal for Mathematical Process and Practice that can be evaluated using a task-specific rubric.

- **Cumulative Assessment** Even-numbered units include this assessment of standards from prior units. All items are appropriate for grading because they match the expectations for progress on those standards at that point in the grade.

- **Challenge** Each Progress Check lesson includes one or more challenge problems related to important ideas from the unit.

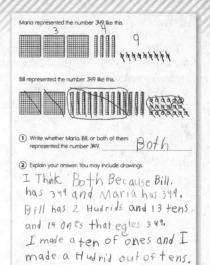

Interim Assessments

These assessments are administered at the beginning, middle, and end of the school year.

- **Beginning-of-Year Assessment** This assessment provides information about children's knowledge and skills related to the content in the first two or three units based on grade-level expectations from the prior grade. It can be useful for RtI screening.

- **Mid-Year and End-of-Year Assessments** These assessments offer teachers snapshots of children's performance on representative samples of standards covered to date. All items are appropriate for grading because they match the expectations for the standards assessed to that point in the year.

Summaries of Assessment Opportunities

Every Unit Organizer includes information about assessment opportunities in the unit. Every Progress Check has tables that summarize the content and process/practice standards assessed in the lesson.

Other Assessment Opportunities

Almost any task in *Everyday Mathematics* can provide information that could be useful for assessment. Assessment tools and systems in *Everyday Mathematics* can accommodate data from sources other than those listed above, and teachers should use their judgment about expanding the range of data they gather and use to inform their instruction and assign grades.

Assessment Tools

Everyday Mathematics provides a variety of tools for collecting, storing, analyzing, reporting, and using assessment data.

Rubrics for Evaluating Mathematical Process and Practice Standards

The Open Response and Reengagement lessons and the Open Response Assessments are opportunities to assess children's progress on the Standards for Mathematical Process and Practice. Each of these includes a task-specific rubric that can be used to evaluate children's work for a specific *Everyday Mathematics* Goal for Mathematical Process and Practice. See sample rubric on the following page.

Sample Rubric

Goal for Mathematical Process and Practice GMP2.2 Make sense of the representations you and others use.	Not Meeting Expectations	Partially Meeting Expectations	Meeting Expectations	Exceeding Expectations
	May state that one or both of the representations are correct, but provides no evidence of interpreting the base-10 blocks as ones, tens, and hundreds.	States that one or both of the representations are correct and provides evidence in words, drawings, or number models of interpreting base-10 blocks as ones, tens, and hundreds, but does not provide evidence of interpreting 10 ones as a ten and 10 tens as a hundred.	States that both of the representations are correct, and provides evidence in words, drawings, or number models of interpreting 10 ones as a ten and 10 tens as a hundred.	Meets expectations and provides evidence in at least two forms (words, drawings, or number models), each of which provides evidence on its own of interpreting 10 ones as a ten and 10 tens as a hundred. Or, meets expectations and refers to Maria's representation as most efficient (e.g., saying that it shows the number with the fewest blocks).

Go Online for generic rubrics for the Goals for Mathematical Process and Practice that you can complete and use to evaluate children's responses to Writing/Reasoning prompts and certain items in the Assessment Check-Ins, Open Response and Reengagement lessons, and Progress Checks. For information on how to use the rubrics, see the Assessment section in the *Implementation Guide*.

Individual Profiles of Progress and Class Checklists

Individual Profiles of Progress (IPPs) combine data from various sources for individual children. Class Checklists facilitate collecting and recording data for an entire class. Blank masters of these forms are provided in this handbook.

Go Online for unit-specific versions of the IPPs and Class Checklists that you can download and print or complete digitally.

Individual Profile of Progress

Class Checklist

Assessment Masters and Answer Keys

This handbook includes masters for the Progress Check assessments and the interim assessments. Answers for the Progress Check assessments are provided at point of use in the *Teacher's Lesson Guide*. Answers for the interim assessments are provided on pages 120–122 of this handbook.

Go Online for answer keys that you can view at full size.

Assessment and Reporting Tools

Digital tools available to teachers through McGraw-Hill's ConnectED platform centralize evaluation, reporting, and targeted differentiation. Teachers are able to evaluate children's work, whether completed digitally or in print, and generate reports on children's progress based on standards covered in lessons and units. The system can track children's performance and provide teachers information and access to materials at point of use to support differentiation decisions.

Unit 1 Self Assessment

Put a check in the box that tells how you do each skill.

Skills	I can do this. I can explain how to do this.	I can do this by myself.	I can do this with help.
① Count coins. MJ1 2			
② Place numbers on a number line. MJ1 1			
③ Skip count. MJ1 9			
④ Use <, >, and =. MJ1 12			
⑤ Add and subtract numbers. MJ1 4			
⑥ Determine odd and even numbers. MJ1 7			

Unit 1 Assessment

(1) Count by ones. Fill in the missing numbers.

    a. 45, _____, _____, 48, _____

    b. 96, 97, _____, 99, _____

    c. 115, 116, _____, _____, _____

(2) If you have 3 dimes, how many cents do you have?

    Total: _____ ¢

(3) Fill in the missing numbers.

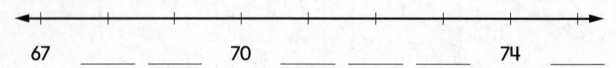

    67   _____   _____   70   _____   _____   _____   74   _____

(4) **a.** Write the number 6 in the correct place on the number line.

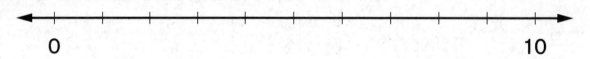

    0   10

    b. Explain how you knew where to write the number.

    _____

    _____

    _____

Unit 1 Assessment (continued)

⑤ Fill in the missing numbers.

23		

		86

⑥ Start at 120 and skip count by 5s. Circle your counts.

									120
121	122	123	124	125	126	127	128	129	130
131	132	133	134	135	136	137	138	139	140

⑦ Write <, >, or =.

> < is less than
> > is more than
> = is the same as

a. 17 _____ 7

b. 25¢ _____ Ⓠ

c. Ⓓ, Ⓓ, Ⓝ, Ⓟ, Ⓟ _____ Ⓠ Ⓝ

d. 37 _____ 73

e. Explain how you know which is the greater number for 7d.

_____

_____

Unit 1 Assessment (continued)

⑧ Beth is playing *Fishing for 10*. She has a 5 in her hand.

    a. What card should she fish for? _____

    b. Complete the number model to show her total after she gets
    the card she fished for.

      5 + _____ = 10

⑨ How many s? Label **odd** or **even**.

    a.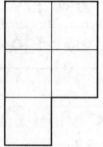

    _____

    _____

    b.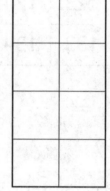

    _____

    _____

Unit 1 Challenge

1 The ② and ⑤ keys on Mike's calculator are broken! How can he still use it to show 25? Use addition and subtraction. Write a number sentence to show what you did.

_____

2 Lila bought an eraser and pencil at the school store. Use Ⓠ, Ⓓ, Ⓝ, and Ⓟ to show how much money she spent in all. Use toolkit coins to help.

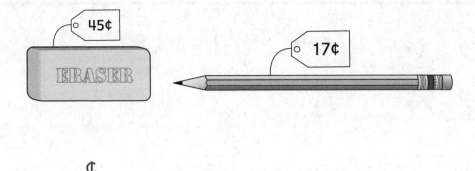

_____ ¢

3 Use Ⓠ, Ⓓ, Ⓝ, and Ⓟ to show 57¢ using as few coins as possible.

Unit 1 Open Response Assessment
Number Lines

1 Place the number 3 in the correct spot on this number line.

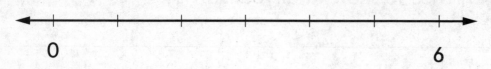

0   6

2 Place the number 3 in the correct spot on this number line.

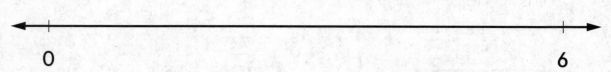

0   6

3 Explain what you did to solve Problem 2.

_____

_____

_____

_____

_____

Unit 2 Self Assessment

Put a check in the box that tells how you do each skill.

Skills	I can do this. I can explain how to do this.	I can do this by myself.	I can do this with help.
① Tell what I saw in a Quick Look.			
② Know doubles facts. **MJ1 22**			
③ Know combinations of 10. **MJ1 22**			
④ Use counters to figure out if a number is odd or even. **MJ1 36**			
⑤ Say two different names for a number. **MJ1 38**			
⑥ Use the turn-around rule. **MJ1 22**			

Unit 2 Assessment

1 Add.

   a. $4 + 4 = $ _____   **d.** $8 + $ _____ $= 10$

   b. $3 + 7 = $ _____   **e.** _____ $+ 1 = 10$

   c. _____ $+ 6 = 12$ **f.** $9 + $ _____ $= 18$

2 Mallory knew that she could use $8 + 2 = 10$ to figure out the sum for $8 + 4$. Explain Mallory's thinking. You may use a double ten frame.

3 For each fact, write a helper fact. You can use a double ten frame to help.

   a. $7 + 8 = ?$ **b.** $7 + 9 = ?$

   Helper fact: Helper fact:

   _____   _____

   $7 + 8 = $ _____ $7 + 9 = $ _____

Unit 2 Assessment (continued)

④

a. Write an addition fact   b. Write the turn-around
    for the domino.   fact.

_____   _____

⑤ Write at least five names in the 12 box.

12

⑥ Take an even number of pennies.

How many pennies did you take? _____

How do you know that the number of pennies is even?

_____

_____

Write a number model with your number of pennies as the sum.
Use equal addends.

_____

Unit 2 Challenge

① Katie knew that 70 + 70 = 140. Explain how she can use this double to solve 70 + 80.

② In the first frame, draw 7 counters one way. In the second ten frame, draw 7 counters a different way.

Below each ten frame, explain how the counters show the number 7. Use numbers or words.

Names for 7.

   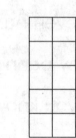

_____   _____

_____   _____

_____   _____

_____   _____

Unit 2 Cumulative Assessment

① Fill in the missing numbers.

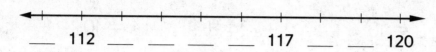

_____ 112 _____ _____ _____ 117 _____ _____ 120

② Count by 5s.

5, _____, _____, _____,

_____, _____, _____,

_____, _____, _____,

_____

③ Count by 10s.

80, _____, _____, _____,

_____, _____, _____,

_____, _____, _____,

_____

④ Write all the double facts.

_____   _____

_____   _____

_____   _____

_____   _____

_____

Unit 2 Cumulative Assessment (continued)

(5) Write all the combinations of 10. Include the turn-around facts.

_____ + _____ = 10   _____ + _____ = 10

_____ + _____ = 10   _____ + _____ = 10

_____ + _____ = 10   _____ + _____ = 10

_____ + _____ = 10   _____ + _____ = 10

_____ + _____ = 10   _____ + _____ = 10

_____ + _____ = 10

(6) How much money?

_____ ¢

Unit 3 Self Assessment

Put a check in the box that tells how you do each skill.

Skills	I can do this. I can explain how to do this.	I can do this by myself.	I can do this with help.
① Write fact families. *MJ1 47*			
② Solve "What's My Rule?" problems. *MJ1 56*			
③ Use strategies to solve subtraction facts. *MJ1 48*			
④ Figure out my number in *Salute!*. *MRB 162–163*			
⑤ Figure out the covered number on a fact triangle. *MRB 47*			
⑥ Solve – 0 and – 1 facts. *MJ1 48*			

Unit 3 Assessment

① Write the fact family.

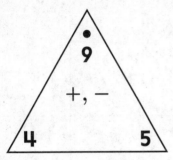

_____ + _____ = _____

_____ + _____ = _____

_____ − _____ = _____

_____ − _____ = _____

② Beth is playing *Salute!*

The dealer says 12.
Her partner has a 5 on his forehead.

What number does Beth have? _____
How do you know? Explain your thinking.

_____

_____

_____

③ José is playing *Subtraction Top-It*. He takes a 16 and a 9.

Write the subtraction fact. _____

④ Solve.

   a. 12 − 3 = _____   **b.** 9 − 7 = _____

   c. Explain how you solved **one** of the facts above.

_____

_____

Unit 3 Assessment (continued)

5. Solve.

Rule
+5

in	out
5	
9	
11	
25	

6. Martin made a 10 to figure out the sum for $8 + 4$. Explain Martin's thinking.

_____

_____

7. Subtract.

a. $5 - 0 =$ _____

b. $7 -$ _____ $= 6$

c. _____ $- 0 = 10$

d. _____ $- 1 = 8$

Unit 3 Challenge

① Zoe wrote the fact family below and said it matches this Fact Triangle.

$3 + 9 = 12$

$9 + 3 = 12$

$9 - 3 = 6$

$9 - 6 = 3$

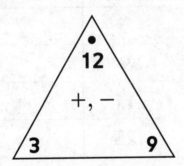

Do you agree with Zoe? _____

Write an argument that explains why you agree or disagree with Zoe.

_____

_____

_____

_____

Unit 3 Challenge (continued)

② **a.** Four children shared their strategies for solving $17 - 8$.
Fill in the circle next to the strategy that **does not** work.

    (A) Jack said, "I know that $8 + 8 = 16$, so I also know that $16 - 8 = 8$. Since 17 is one more than 16, $17 - 8 = 9$."

    (B) Susan said, "I counted up 2 from 8 to 10, and then I counted up 7 more to get 9."

    (C) Jessica said, "I subtracted 7 from 17 to get 10. Then I subtracted 1 more to get to 9."

    (D) Angel said, "I pointed at 8 and said '1' and then counted up until I got to 17. I counted 10."

    b. What must the child do differently so that the strategy **does** work?

_____

_____

_____

Unit 3 Open Response Assessment
A Subtraction Strategy

Grace solved 12 − 7 this way:

"I started at 12 and took away 2 to get to 10. Then I took away 5 more. I ended up at 5. So, 12 − 7 = 5."

Grace solved 13 − 4 this way:

"I started with 13 and took away 3 to get to 10. Then I took away 1 more. I ended up at 9. So, 13 − 4 = 9."

Show and explain how to use Grace's subtraction strategy to solve 14 − 8.

Unit 4 Self Assessment

Put a check in the box that tells how you do each skill.

Skills	I can do this. I can explain how to do this.	I can do this by myself.	I can do this with help.
① Tell time. MJ1 67			
② Write time. MJ1 67			
③ Know A.M. and P.M. MJ1 71			
④ Write a number represented by base-10 blocks. MB 86			
⑤ Write a number in expanded notation. MJ1 74			
⑥ Measure lengths. MJ1 87			

Unit 4 Assessment

① Write the time.

a.    b.    c.

_____:_____   _____:_____   _____:_____

② Draw the hands to match the time.

a. 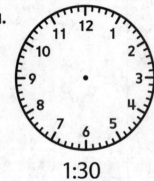   b.

   1:30   6:00

③ Write A.M. or P.M. to make the sentence true.

a. School starts at 9:00 _____

b. I do my homework at 4:30 _____

c. I read a bedtime story at 8:00 _____

d. I go to sleep for the night at 8:30 _____

Unit 4 Assessment (continued)

④ **a.** What number do the blocks show? _____

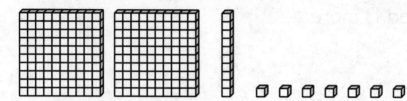

b. Write the number using expanded notation.

_____

⑤ **a.** Write the numbers using expanded notation.

52 = _____

25 = _____

b. Write < or >.

52 _____ 25

c. Explain how you knew which number was larger.

_____

_____

_____

Unit 4 Assessment (continued)

⑥ Use base-10 shorthand.

    a. Show 29. Add 13 more.

    b. How many in all? _____

    c. Show the total with the fewest possible blocks.

Measure each line segment to the nearest inch and nearest centimeter.

⑦ _____

    a. about _____ inches   **b.** about _____ centimeters

⑧ _____

    a. about _____ inches   **b.** about _____ centimeters

Unit 4 Challenge

① Victor says he can use the ruler below to measure the length of his crayon.

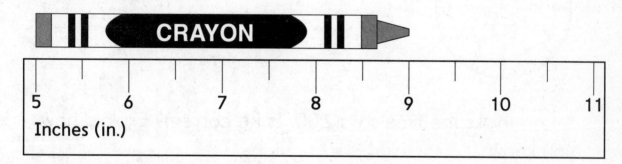

Rosa says that Victor's ruler is broken and says that he has to use her ruler.

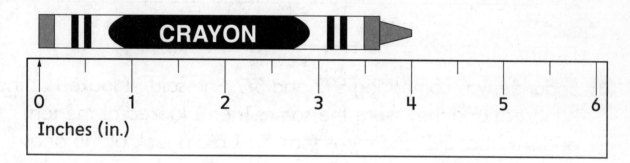

Can Victor use his ruler to measure the crayon? _____
Explain your answer.

_____

_____

_____

Unit 4 Challenge (continued)

(2)

Simon thinks the time says 2:00. Is he correct? Explain how you know.

_____

_____

_____

(3) Shaunee was comparing 592 and 577. She said, "I looked at the hundreds and they were the same. Then I looked at the tens and saw that 592 had more tens. So I didn't look at the ones because I already knew 592 was larger."

Do you agree that Shaunee didn't have to look at the ones? Explain your answer.

_____

_____

_____

Unit 4 Cumulative Assessment

(1) Write the time.

_____ : _____   _____ : _____

(2) Solve.

   a. $0 + 9 =$ _____   **b.** $5 +$ _____ $= 5$ **c.** $7 - 0 =$ _____

   d. For Problems 2a–2c, what patterns do you notice?

_____

_____

(3) Solve.

   a. $1 + 6 =$ _____   **b.** $4 +$ _____ $= 5$ **c.** $8 - 1 =$ _____

   d. For Problems 3a–3b, what patterns do you notice?

_____

_____

(4) Fill in the missing numbers.

0   25   _____   75   _____   _____   175

Unit 4 Cumulative Assessment (continued)

⑤ If you have 6 nickels, how many cents do you have?

Total: _____

⑥ If you have 6 quarters, how many cents do you have?

Total: _____

⑦ Fill in the missing numbers.

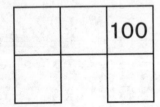

		100

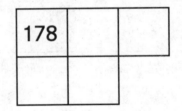

178		

⑧ ☆ ☆ ☆ ☆ ☆ ☆ ☆ ☆
☆ ☆ ☆ ☆ ☆ ☆ ☆ ☆

a. Is the number of stars odd or even? _____

b. Explain. _____
_____

c. Write a number model with the number of stars as the sum. Use equal addends. _____

Unit 5 Self Assessment

Put a check in the box that tells how you do each skill.

Skills	I can do this. I can explain how to do this.	I can do this by myself.	I can do this with help.
① Count coins. MJ2 107			
② Make change. MJ2 109			
③ Solve number stories. MJ2 120, 121, 123, 125, 126, 130			
④ Add and subtract 10. MJ2 130 Problem 2			
⑤ Add and subtract 100. MJ2 130 Problem 2			
⑥ Use an open number line. MJ2 117			

Unit 5 Assessment

(1) Write + or − to make the number sentence true.

    a. $10 = 6$ _____ 4

    b. 7 _____ $5 = 2$

    c. $11 = 8$ _____ 3

In Problems 2–4, use Ⓟ, Ⓝ, Ⓓ, and Ⓠ to show your answer.

(2) You buy a green pepper for 27¢. Show the coins you could use to pay the exact amount.

(3) You buy some lettuce for 45¢. Show the coins you could use to pay the exact amount.

(4) You buy yogurt for 70¢. Show the coins you could use to pay the exact amount.

(5) Solve.

    a. $57 + 10 =$ _____   **b.** $94 - 10 =$ _____

    c. $98 + 10 =$ _____   **d.** _____ $= 100 - 10$

    e. $120 + 10 =$ _____   **f.** _____ $= 400 + 100$

    g. $200 - 100 =$ _____   **h.** _____ $= 130 - 10$

Unit 5 Assessment (continued)

(6) Use an open number line to help you solve the story.

Mrs. Peters had 22 pencils to give to her students. She bought 35 more. How many does she have now? _____ pencils

⟵――――――――――――――――――――――⟶

In Problems 7–9, write a number model for each problem. You may use the diagrams to help. Then solve.

(7) Wade put 7 blueberries on his cereal. Then he added 9 more. How many does he have now?

Number model: _____

Answer: _____ blueberries

Start	Change	End
	_____	

(8) 8 bunnies were sitting on the grass. 6 more joined them. How many bunnies are there altogether?

Number model: _____

Answer: _____ bunnies

Total	
Part	Part

(9) The temperature was 60 degrees in the afternoon. It was 40 degrees in the evening. How much did the temperature change?

Number model: _____

Answer: _____ degrees

Start	Change	End
60	−?	40

Unit 5 Challenge

① Carlos buys a hamburger for 65¢ and gives the clerk $1. The clerk gives him a quarter for change. Carlos says the amount of change is incorrect. Do you agree with Carlos? Explain your answer.

_____

_____

_____

_____

② Dana said that $192 + 10 = 102$. Explain Dana's mistake.

_____

_____

_____

_____

_____

_____

Unit 5 Open Response Assessment
Buying from a Vending Machine

Carlos wants to buy chocolate milk from the vending machine. The milk costs 75¢. Carlos has 2 quarters, 5 dimes, and 5 nickels.

(1) Show at least four possible coin combinations Carlos could use to pay for the milk. Use Ⓝ, Ⓓ, and Ⓠ to record your answers.

Unit 5 Open Response Assessment (continued)

② Pick one of your coin combinations and show or explain how you know it totals exactly 75¢.

Unit 6 Self Assessment

Put a check in the box that tells how you do each skill.

Skills	I can do this. I can explain how to do this.	I can do this by myself.	I can do this with help.
1 Answer questions using data on a picture graph. MJ2 134-135			
2 Find the difference between two amounts in a number story. MJ2 137			
3 Solve two-step number stories. MJ2 150-151			
4 Solve number stories about lengths and heights of animals. MJ2 147			
5 Solve addition problems and record what I did. MJ2 153			
6 Use partial-sums addition to add multidigit numbers. MJ2 155, 158			

Unit 6 Assessment

① Use the picture graph below to answer the questions.

How Many Pencils?

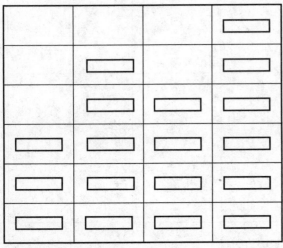

Mateo Elena Maria David

KEY: each ⬜ = 1 pencil

Who has the most pencils? _____

Who has the fewest pencils? _____

How many more pencils does David have than Maria? _____

Unit 6 Assessment (continued)

For Problems 2–5:

- Write a number model with a ? to show what you need to find.
  You may draw a [Total / Part | Part], [Start ⌒Change End ___], or [Quantity / Quantity ___ Difference] to help.

- Solve the problem.

- Write the answer.

② Fish E is 40 inches long. Fish F is 30 inches long.

How much longer is Fish E than Fish F?

Number model: _____

Fish E is _____ inches longer than Fish F.

③ Connor needed 20 juice boxes for her class party. She bought 22. How many extra juice boxes did she have?

Number model: _____

Answer: _____ juice boxes

Unit 6 Assessment (continued)

④ The giraffe is 13 feet tall when standing.
Its legs are 6 feet long.

How tall is the giraffe when it is lying down?

Number model: _____

Answer: _____ feet

⑤ The green ribbon is 15 feet long. The white
ribbon is 10 feet long. Which is longer,
the green ribbon or the white ribbon?

_____

How much longer?

Number model: _____

Answer: _____ feet

Unit 6 Assessment (continued)

⑥ Solve.

Elmer had 6 yards of rope.
He found 8 more yards in the garage.
He cut off 5 yards for his project.
How many yards does he have now?

Answer: _____ yards

⑦ Make a ballpark estimate and then choose any method to solve.
Use your estimate to check if your answer makes sense.

Show your work.

39 + 46

Ballpark estimate: _____

39 + 46 = _____

⑧ Solve using partial-sums addition. Show your work.
You may use base-10 blocks to help. If you use blocks,
record your work in base-10 shorthand.

$$\begin{array}{r} 26 \\ + \ 42 \\ \hline \end{array}$$
    $$\begin{array}{r} 194 \\ + \ 235 \\ \hline \end{array}$$

Unit 6 Challenge

① Addie gathered 5 eggs in the hen house. 3 eggs fell out of her basket and broke. She then found 12 more eggs. How many eggs does Addie have now?

Jim wrote these number models to help him solve this problem.

$5 + 3 = ?$
$8 + 12 = ?$

Is he correct? Explain.

_____

_____

_____

② Melissa and Ebe both solved the same problem using partial sums. Their work is shown below.

Melissa	Ebe
156	156
+ 338	+ 338
400	14
80	80
14	400
494	494

Explain why both strategies work.

_____

_____

_____

Unit 6 Cumulative Assessment

1 Write each number in expanded form.

    a. 87 _____

    b. 134 _____

    c. 678 _____

    d. 810 _____

2 Write the number that has

    a. 7 in the 100s place,
       0 in the 10s place,
       0 in the 1s place. _____

    b. 5 in the 100s place,
       6 in the 10s place,
       4 in the 1s place. _____

    c. 4 in the 10s place,
       2 in the 100s place,
       1 in the 1s place. _____

3   **a.** Justin has Ⓓ Ⓓ Ⓝ Ⓝ.
       How much money
       does he have? _____

    b. Javier has Ⓝ Ⓝ Ⓝ Ⓝ
       Ⓓ Ⓓ Ⓓ Ⓟ Ⓟ Ⓟ Ⓟ Ⓟ Ⓟ.
       How much money
       does he have? _____

4 Use <, >, or =.

    a. 133 _____ 233

    b. 241 _____ 246

    c. 891 _____ 891

    d. 656 _____ 636

5 Katy is playing *Exchange Game* with Base-10 blocks. Below are the blocks she has. Circle the longs she needs to trade.

Unit 6 Cumulative Assessment (continued)

⑥ Measure the line segments to the nearest inch and centimeter.

a. _____

about _____ inches

about _____ centimeters

b. _____

about _____ inches

about _____ centimeters

c. Explain why there are more centimeters than inches in the measures for Problem 6b.

_____

_____

_____

⑦ Solve.

a. $158 + 10 =$ _____

b. _____ $= 10 + 553$

c. $100 + 367 =$ _____

d. $200 +$ _____ $= 300$

e. _____ $+ 112 = 122$

f. _____ $= 627 + 100$

g. $233 - 10 =$ _____

h. $600 - 100 =$ _____

i. $341 -$ _____ $= 241$

j. $178 -$ _____ $= 168$

Unit 6 Cumulative Assessment (continued)

⑧ Match the number to the number word.

26   One hundred forty-six

33   Thirty-three

146   Six hundred

600   Twenty-six

⑨ Use words to write the number name.

a. 500 _____

b. 526 _____

⑩ Alex says that 700 is 7 + 0 + 0 in expanded form.

a. Do you agree with Alex? _____

b. Why? _____

⑪ Justin has 30¢ in his coat pocket.

a. What coins might he have? Show your answer using Ⓠ, Ⓓ, Ⓝ, and/or Ⓟ.

b. Show a different combination of coins Justin might have.

Unit 7 Self Assessment

Put a check in the box that tells how you do each skill.

Skills	I can do this. I can explain how to do this.	I can do this by myself.	I can do this with help.
① Play *Hit the Target*. MJ2 165			
② Add 3 or more numbers. MJ2 171			
③ Measure objects to the nearest inch and centimeter. MJ2 180-181			
④ Complete a line plot. MJ2 187			
⑤ Use personal references to help estimate length. MJ2 175			
⑥ Use measuring tools correctly. MJ2 178			

Unit 7 Assessment

① Fill in the missing numbers.

a. $33 + \underline{\hspace{2cm}} = 40$ **b.** $90 = 88 + \underline{\hspace{2cm}}$

c. $\underline{\hspace{2cm}} + 45 = 50$ **d.** $30 = \underline{\hspace{2cm}} + 24$

② Two teams are playing *Basketball Addition*. Here are their scores for the first half:

Team A	Team B
Player 1: 10	Player 1: 20
Player 2: 12	Player 2: 5
Player 3: 6	Player 3: 15
Player 4: 8	Player 4: 3

a. Find each team's score for the first half.

Team A: $\underline{\hspace{1.5cm}}$ points   Team B: $\underline{\hspace{1.5cm}}$ points

b. Which team is winning at halftime? $\underline{\hspace{5cm}}$

③ **a.** What measuring tool would you use to measure the length of your Math Journal? $\underline{\hspace{5cm}}$

b. Explain your choice. $\underline{\hspace{6cm}}$

$\underline{\hspace{8cm}}$

c. What unit would you use when measuring the length of your Math Journal? $\underline{\hspace{5cm}}$

d. Explain your choice. $\underline{\hspace{6cm}}$

$\underline{\hspace{8cm}}$

Unit 7 Assessment (continued)

④ Estimate and then measure the length of each picture below in inches and centimeters.

a.

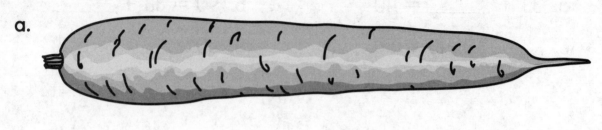

Estimates:

about _____ inches

about _____ cm

Measures:

about _____ inches

about _____ cm

b.

Estimates:

about _____ inches

about _____ cm

Measures:

about _____ inches

about _____ cm

c. Did you get a smaller number when you measured in inches or centimeters? _____ Why?

_____

_____

Unit 7 Assessment (continued)

⑤ For each unit write the name of something you might use as a personal reference.

a. centimeter _____

b. inch _____

c. yard _____

d. meter _____

⑥ Use the information in the tally chart to complete the line plot.

Room 102 Head Sizes		
Head Sizes (cm)	Number of Children	
	Tallies	Number
46	//	2
47		0
48	////	4
49	~~////~~/	5
50	/	1
51	///	3
52	/	1
Total =		16

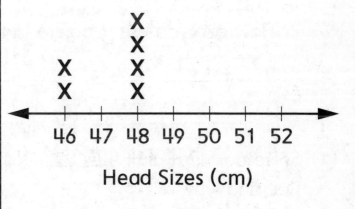

Room 102 Head Sizes (cm)

Head Sizes (cm)

Unit 7 Challenge

① You are playing *Hit the Target*. Your target number is 70 and your starting number is 36.

a. Show how you can hit the target number in two changes.

Target number: __70__

Starting number	Change	Result	Change	Result
36				

b. Show how you can hit the target number in one change.

Target number: __70__

Starting number	Change	Result
36		

c. How can you use Problem 1a to help you solve Problem 1b?

_____

_____

② Solve $6 + 19 + 11 + 4.$ _____ Explain which numbers you added first and why.

_____

_____

Unit 7 Open Response Assessment
Using Base-10 to Show Numbers

Maria represented the number 349 like this.

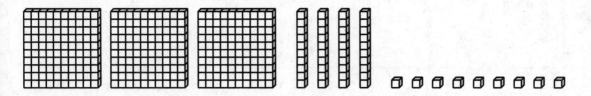

Bill represented the number 349 like this.

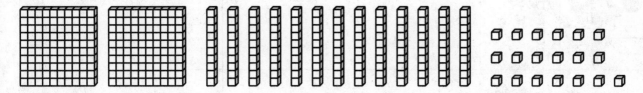

① Write whether Maria, Bill, or both of them represented the number 349.

_____

② Explain your answer. You may include drawings.

Unit 8 Self Assessment

Put a check in the box that tells how you do each skill.

Skills	I can do this. I can explain how to do this.	I can do this by myself.	I can do this with help.
1 Find shapes that have a given number of sides and angles. *Game — Shape Capture*			
2 Name shapes (triangles, quadrilaterals, pentagons, and hexagons). *MJ2 197*			
3 Partition a rectangle into rows and columns. *MJ2 207*			
4 Find the total number of objects in an array. *MJ2 212*			
5 Write an addition number model for an array. *MJ2 212*			
6 Solve an array number story. *MJ2 210*			

Unit 8 Assessment

① Draw a 3-sided shape.

② Draw a 4-sided shape with 4 right angles.

What is the name of your shape?

_____

What is the name of your shape?

_____

③ Circle the shapes that have parallel sides.

④ Describe the shape below. Use the words *sides* and *angles*. Then write the name of the shape.

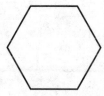

_____

_____

Shape name: _____

Unit 8 Assessment (continued)

⑤ Your teacher will give you a shape. Describe the shape and write its name.

_____

_____

⑥ Partition the rectangle into 2 rows with 2 same-size squares in each row.

How many squares did you draw? _____

Unit 8 Assessment (continued)

⑦ Allen set up 4 rows of chairs with 3 chairs in each row.

a. Draw an array to show how Allen set up the chairs.

b. How many chairs did Allen set up? _____ chairs

c. Write a number model for the array.

_____

Unit 8 Challenge

①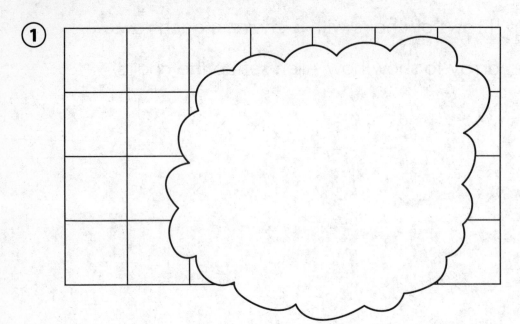

How many rows of squares are on the rectangle? _____

How many squares are in each row? _____

How many squares are there in all? _____

② A store clerk had 20 boxes of cereal to put on display. She arranged them in rows. She put 4 in each row.

a. Draw the rows of cereal boxes.

b. How many rows were in the display? _____ rows

c. Write an addition number model for the array.

_____

Unit 8 Cumulative Assessment

(1) Write the time shown on each clock.

a.

b.

c.

_____   _____   _____

(2) Rachel estimated the length of each line segment.

a. _____

about ___6___ inches

Do you agree with her estimate? _____ Why?

_____

_____

How long do you think the line segment is?

About _____ inches

b. _____

about _4 or 5_ centimeters

Do you agree with her estimate? _____ Why?

_____

_____

How long do you think the line segment is?

About _____ centimeters

Unit 8 Cumulative Assessment (continued)

③ Solve. Try to make friendly numbers.

$23 + 17 + 10 + 12 =$ _____

$16 + 31 + 14 + 19 =$ _____

Pick one of the problems above. Explain how you added the numbers.

_____

_____

_____

_____

_____

_____

_____

_____

_____

_____

_____

Unit 8 Cumulative Assessment (continued)

④ For each problem:

- Write a number model with a ? to show what you need to find.

- To help you may draw a

Total	
Part	Part

, Start ⌢Change End _____ , or

Quantity	
Quantity	
	Difference

.

- Solve the problem and write the answer.

a. The Giant Squid is 55 feet long and the Saltwater Crocodile is 23 feet long. How long are they altogether?

23 feet   55 feet

Number model: _____

Answer: _____ feet

b. The Blue Whale is 98 feet long. Together, the Blue Whale and the Green Anaconda are 124 feet long. How long is the Green Anaconda?

Number model: _____

Answer: _____ feet

Unit 8 Cumulative Assessment (continued)

⑤ The gym at McKenzie School has a basket of sports balls. Curtis sorted the balls. He made the tally chart below.

Type of Ball	Tallies	Number
Basketball	~~HHT~~ ////	9
Soccer Ball	~~HHT~~ /	6
Football	////	4
Softball	~~HHT~~	5

Draw a picture graph to show his data.

Sports Balls at McKenzie School

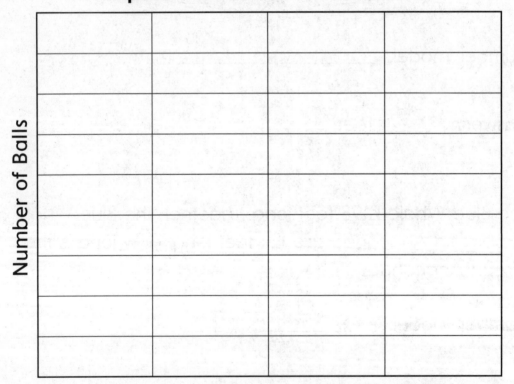

Number of Balls

Type of Balls

KEY: each ◯ = 1 ball

Unit 8 Cumulative Assessment (continued)

⑥ Alison's class graphed the weather data for the month of February.

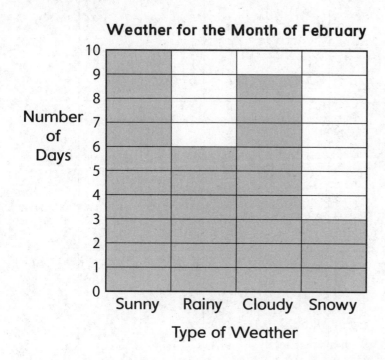

Weather for the Month of February

Number of Days

Sunny   Rainy   Cloudy   Snowy

Type of Weather

Use the graph to answer the questions.

How many more sunny days were there than snowy days? _____

How many days were either cloudy or rainy? _____

How many days in February were recorded on this graph? _____

Write a question that can be answered with this graph.

_____

Write the answer to your question.

_____

Unit 9 Self Assessment

Put a check in the box that tells how you do each skill.

Skills	I can do this. I can explain how to do this.	I can do this by myself.	I can do this with help.
① Divide shapes into equal shares. **MJ2** 220-221			
② Measure the lengths of objects. **MJ2** 227			
③ Write 3-digit numbers in expanded form. **MJ2** 230			
④ Use <, >, or = to compare 3-digit numbers. **MRB** 74-75			
⑤ Solve number stories about equal groups. **MJ2** 242			
⑥ Solve 2-digit subtraction problems. **MJ2** 234			

Unit 9 Assessment

① **a.** Divide this shape into fourths.

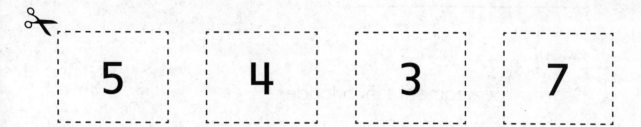

   b. Name one part.

_____

   c. Name all the parts.

_____

② **a.** Divide this shape into thirds.

   b. Name one part.

_____

   c. Name all the parts.

_____

Unit 9 Assessment (continued)

3 **a.** Measure the length of the line segment to the nearest inch.

_____

about _____ inches

b. Draw a line segment 1 inch longer.

My line segment is about _____ inches long.

4 **a.** Write each number in words.

413 _____

431 _____

b. Write each number in expanded form.

413 = _____

431 = _____

c. Write <, >, or = to compare the numbers.

413 ☐ 431

5 Brenda put her stuffed animals in 2 rows on her bed. There were 8 stuffed animals in each row. How many stuffed animals were there in all?

_____ stuffed animals

Addition number model: _____

Unit 9 Assessment (continued)

⑥ 10 children divide into 2 teams to play kickball. Each team has an equal number of children. How many children are on each team?

_____ children

Addition number model: _____

⑦ Write a number sentence to show a ballpark estimate and then solve. Show your work.

Unit

a. $34 - 17 = ?$

Ballpark estimate:

_____

Solution:

Answer: _____

b. $127 - 44 = ?$

Ballpark estimate:

_____

Solution:

Answer: _____

⑧ a. How much money is 10 nickels? _____ cents

b. How many is 10 [5s]? _____

⑨ a. How much money is 7 dimes? _____ cents

b. How many is 7 [10s]? _____

Unit 9 Assessment (continued)

⑩ Use Ⓠ Ⓓ Ⓝ Ⓟ to show at least one way to pay for the orange juice.

Orange Juice
8 oz
89¢

One Way　　　　　　**Another Way**

⑪ Divide the shape into thirds.

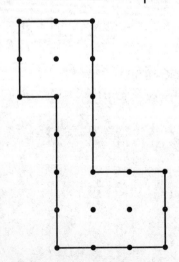

Unit 9 Challenge

① **a.** Claude has 2 rows of pea plants in his garden. There are 7 pea plants in each row. How many pea plants are in the garden?

pea plants _____

b. Write an addition number model and a multiplication number model for the story in Part a.

Addition number model: _____

Multiplication number model: _____

c. How did you know which numbers to write in your addition number model?

_____

_____

How did you know which numbers to write in your multiplication number model?

_____

_____

② Divide the shape into 4 equal parts in two different ways.

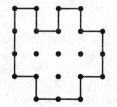   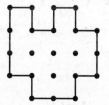

Unit 9 Open Response Assessment
Finding the Largest Sum

Cut out the 4 digits from the bottom of this page.

Make two 2-digit numbers in the boxes below so that when you add them you get the largest possible sum.

Use the digits you cut out to help you try different combinations. When you find the combination that makes the largest sum, write the numbers in the boxes.

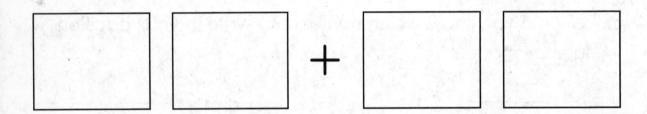

The largest sum is _____.

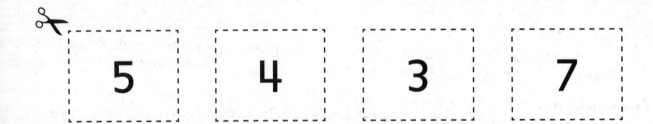

Unit 9 Open Response Assessment (continued)

Explain how you know you found the largest sum.

Beginning-of-Year Assessment

The Beginning-of-Year Assessment can be used to gauge children's readiness for the content they will encounter in second grade.

Goals

The following table provides information about the standards and the Standards for Mathematical Process and Practice assessed in the Beginning-of-Year Assessment.

Standards	Goals for Mathematical Content (GMC)	Item(s)
2.OA.1	Use addition and subtraction to solve 1-step number stories.	12, 13
	Model 1-step problems involving addition and subtraction.	12, 13
2.OA.2	Add within 20 fluently.	9a, 9c, 9d, 9e, 10–12
	Subtract within 20 fluently.	9b, 9f, 10–11, 13
	Know all sums of two 1-digit numbers automatically.	9a, 9c, 9d, 9e, 10a, 10b, 11–13
2.NBT.2	Count by 1s.	1, 6, 7
	Counts by 5s, 10s, and 100s.	2, 3, 6–8
2.NBT.4	Record comparisons using >, =, or <.	10
2.NBT.8	Mentally add 10 to and subtract 10 from a given number.	6, 7
2.MD.7	Tell and write time using analog and digital clocks.	14a, 14b
2.MD.8	Solve problems involving coins and bills.	4, 5
	Read and write monetary amounts.	4, 5
	Goals for Mathematical Process and Practice (GMP)	
SMP1	Make sense of your problem. **GMP1.1**	11–13
	Solve problems in more than one way. **GMP1.5**	11
SMP7	Look for mathematical structures such as categories, patterns, and properties. **GMP7.1**	6, 7
	Use structures to solve problems and answer questions. **GMP7.2**	6–8

Beginning-of-Year Assessment

Fill in the missing numbers.

(1) 28, 29, _____, _____, 32, _____, 34

(2) 55, 60, _____, 70, _____, 80, _____, _____

(3) _____, _____, 95, 105, _____, 125

Write the amount.

(4)

_____ ¢

(5)

_____ ¢

Beginning-of-Year Assessment (continued)

Write the missing numbers.

(6)

	7		9
	17		
		28	

(7)

41	42	43	44
			54
		63	

(8) Fill in the empty frames.

Rule
Add 10

40 → ☐ → ☐ → 70 → ☐ → 90 → ☐

Beginning-of-Year Assessment (continued)

9 Solve.

   a. $7 + 1 =$ _____

   b. _____ $= 10 - 5$

   c. $6 + 4 =$ _____

   d. $6 + 6 =$ _____

   e. $3 +$ _____ $= 5$

   f. $6 - 1 =$ _____

Unit
pencils

10 Use <, >, or =.

   a. $8 + 2$ _____ $3 + 7$

   b. $4 + 4$ _____ 9

   c. $14 - 7$ _____ $12 - 9$

Unit
crayons

11 Write at least five names for 10 in the box.

10

Beginning-of-Year Assessment (continued)

Solve. Write a number model.

(12) Jace had 4 ice cubes in his cup. He added 3 more.

How many ice cubes are in his cup now? _____ ice cubes

Number model: _____

(13) Dayton had 16 carrots. He gave 8 carrots to his sister.

How many carrots did he have left? _____ carrots

Number model: _____

(14) Write the time.

a.

b.

_____:_____   _____:_____

Mid-Year Assessment

The Mid-Year Assessment covers some of the important concepts and skills presented in *Second Grade Everyday Mathematics*. It should be used to complement the ongoing and periodic assessments that appear within lessons and at the end of each unit.

Goals

The following table provides information about the standards and the Standards for Mathematical Process and Practice assessed in the Mid-Year Assessment.

Standards	Goals for Mathematical Content (GMC)	Item(s)
2.OA.2	Add within 20 fluently.	1a, 1b, 2a–2d, 3, 4
	Subtract within 20 fluently.	1c, 1d, 3
	Know all sums of two 1-digit numbers automatically.	1a, 1b, 2a–2d, 3, 4
2.OA.3	Determine whether the number of objects in a group is odd or even.	5a
	Express an even number as a sum of two equal addends.	5c
2.NBT.2	Count by 1s.	6a, 8a, 8b
	Count by 5s, 10s, and 100s.	7, 8a, 8b
2.NBT.3	Read and write number names.	9a, 9b
2.NBT.8	Mentally add 10 to and subtract 10 from a given number.	8a, 8b, 10
2.MD.6	Represent whole numbers as lengths from 0 on a number-line diagram.	7
2.MD.7	Tell and write time using analog and digital clocks.	13, 14
	Use A.M. and P.M.	14
2.MD.8	Solve problems involving coins and bills.	11, 12
	Read and write monetary amounts.	11, 12

Goals (continued)

Standards	Goals for Mathematical Process and Practice (GMP)	Item(s)
SMP1	Solve problems in more than one way. **GMP1.5**	11
SMP2	Create mathematical representations using numbers, words, pictures, symbols, gestures, tables, graphs, and concrete objects. **GMP2.1**	11, 12
	Make sense of representations you and others use. **GMP2.2**	7, 8c
	Make connections between representations. **GMP2.3**	11, 12
SMP3	Make mathematical conjectures and arguments. **SMP3.1**	13
	Make sense of others' mathematical thinking. **SMP3.2**	13
SMP4	Model real-world situations using graphs, drawings, tables, symbols, numbers, diagrams, and other representations. **GMP4.1**	11–13
	Use mathematical models to solve problems and answer questions. **GMP4.2**	11
SMP6	Explain your mathematical thinking clearly and precisely. **GMP6.1**	5b
SMP7	Look for mathematical structures such as categories, patterns, and properties. **GMP7.1**	2e, 6b, 7
	Use structures to solve problems and answer questions. **GMP7.2**	6b, 7, 8c, 10

Mid-Year Assessment

1. Solve.

   a. $1 + 5 =$ _____

   b. $9 +$ _____ $= 10$

   c. $4 - 1 =$ _____

   d. $7 -$ _____ $= 6$

2. Solve.

   a. $7 + 7 =$ _____

   b. $2 + 9 =$ _____

   c. $10 +$ _____ $= 20$

   d. $8 + 8 =$ _____

   e. For Problems 2a–2d, which fact does not fit the pattern?

   _____

   How is it different?

   _____

   _____

3. Fill in the missing numbers to complete the fact family.

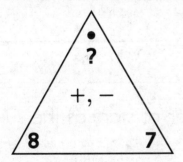

   $8 +$ _____ $=$ _____

   _____ $+$ _____ $=$ _____

   _____ $- 7 =$ _____

   _____ $-$ _____ $=$ _____

Mid-Year Assessment (continued)

(4) Write all the combinations of 10. Include the turn-around facts.

_____ + _____ = 10   _____ + _____ = 10

_____ + _____ = 10   _____ + _____ = 10

_____ + _____ = 10   _____ + _____ = 10

_____ + _____ = 10   _____ + _____ = 10

_____ + _____ = 10   _____ + _____ = 10

_____ + _____ = 10

(5) a. Is the number of stars odd or even?_____

b. Explain your answer.

_____

_____

_____

c. Write a number model with the number of stars as the sum.

_____

Mid-Year Assessment (continued)

6 Fill in the missing numbers.

a.

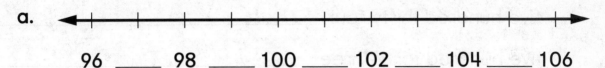

96 _____ 98 _____ 100 _____ 102 _____ 104 _____ 106

b. What is the same about all of the numbers that were missing?

_____

_____

7 Place the number 10 in the correct spot on this number line.

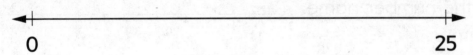

0   25

8 Fill in the missing numbers.

a.

47	48	49
	58	
	68	

b.

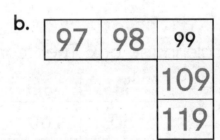

97	98	99
		109
		119

c. Explain how you knew what numbers to write in the boxes for problem 8a.

_____

_____

_____

Mid-Year Assessment (continued)

9 **a.** For each number word write the number.

five hundred thirty-four _____

five hundred forty-three _____

b. Write a number that comes between five hundred thirty-four and five hundred forty-three. _____

Write the number name. _____

10 Write the rule in the box. Then complete the table.

in
↓

Rule

↓
out

in	out
50	60
90	100
	110
120	
	230

Mid-Year Assessment (continued)

(11) Bryn has the following coins:

Ⓟ Ⓟ Ⓟ Ⓟ Ⓟ Ⓟ Ⓟ Ⓟ Ⓟ Ⓟ Ⓝ Ⓝ Ⓝ Ⓓ

How much money does Bryn have? _____

Use Ⓓ, Ⓝ, and Ⓟ to show this amount with fewer coins.

Show the same amount using a different combination of coins.

(12) Evan has 44¢ in his piggy bank.

Use Ⓓ, Ⓝ, and Ⓟ to show the coins he might have in his piggy bank.

Mid-Year Assessment (continued)

13 Farid says the clock at the right reads 9:15.
Yasmin says the clock reads 3:45.

Who is correct? _____

How do you know?

_____

_____

_____

Explain the other child's mistake.

_____

_____

14 Kennedy's swim lesson is at 4:45 P.M.

a. Does she swim in the morning or in the afternoon?

_____

b. Show 4:45 P.M. on the clocks.

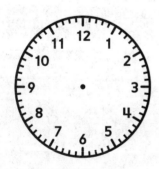
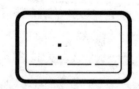

End-of-Year Assessment

The End-of-Year Assessment covers some of the important concepts and skills presented in *Second Grade Everyday Mathematics.* It should be used to complement the ongoing and periodic assessments that appear within lessons and at the end of units.

Goals

The following table provides information about the standards and the Standards for Mathematical Process and Practice assessed in the End-of-Year Assessment.

Standards	Goals for Mathematical Content (GMC)	Item(s)
2.OA.1	Model 1-step problems involving addition and subtraction.	1, 2
	Use addition and subtraction to solve 1-step number stories.	1, 2
	Model 2-step problems involving addition and subtraction.	3, 4
	Use addition and subtraction to solve 2-step number stories.	3, 4
2.OA.4	Find the total number of objects in a rectangular array.	23a
	Express the number of objects in an array as a sum of equal addends.	23b
2.NBT.1	Understand 3-digit place value.	6b, 8
	Represent whole numbers as hundreds, tens, and ones.	6b, 7b
2.NBT.1a	Understand exchanging tens and hundreds.	7b
2.NBT.1b	Understand 100, 200, . . . , 900 as some hundreds, no tens, and no ones.	8
2.NBT.3	Read and write numbers.	6a, 7a
	Read and write numbers in expanded form.	12
2.NBT.4	Compare and order numbers.	9a–9d, 11a, 11c
	Record comparisons using >, =, or <.	9a–9d
2.NBT.5	Add within 100 fluently.	10a
	Subtract within 100 fluently.	10c
2.NBT.6	Add up to four 2-digit numbers.	10a, 10b, 10e, 10f
2.NBT.7	Add multidigit numbers using models or strategies.	10a, 10b, 10e, 10f, 16
	Subtract multidigit numbers using models or strategies.	10c, 10d
2.NBT.9	Explain why addition and subtraction strategies work.	10g

Goals (continued)

2.MD.1	Measure the length of an object.	13c, 13d, 14a, 14b
	Select appropriate tools to measure length.	25
2.MD.2	Measure an object using 2 different units of length.	13c, 13d
	Describe how length measurements relate to the size of the unit.	13e
2.MD.3	Estimate lengths.	13a, 13b
2.MD.4	Measure to determine how much longer one object is than another.	14b, 14c
2.MD.5	Solve number stories involving length by adding and subtracting.	15
	Model number stories involving length.	15
2.MD.6	Represent sums and differences on a number-line diagram.	16
2.MD.7	Tell and write time (to the nearest 5 minutes) using analog and digital clocks.	17a, 17b
2.MD.8	Solve problems involving coins and bills.	18
	Read and write monetary amounts.	18
2.MD.9	Represent measurement data on a line plot.	19a
2.MD.10	Organize and represent data on bar and picture graphs.	20a
	Answer questions using information in graphs.	19b, 19c, 20b–20f
2.G.1	Recognize and draw shapes with specified attributes.	21a, 21b
	Identify 2- and 3-dimensional shapes.	21a
2.G.2	Partition a rectangle into rows and columns of same-size squares and count to find the total number of squares.	22a, 22b
2.G.3	Partition shapes into equal shares.	5a, 24a
	Describe equal shares using fraction words.	5b
	Describe the whole as a number of shares.	5c
	Recognize that equal shares of a shape need not have the same shape.	24a, 24b

Goals (continued)

Standards	Goals for Mathematical Process and Practice (GMP)	Item(s)
SMP1	Check whether your answer makes sense. `GMP1.4`	22c
	Solve problems in more than one way. `GMP1.5`	7a–7b, 18
	Compare the strategies you and others use. `GMP1.6`	7c
SMP2	Create mathematical representations using numbers, words, pictures, symbols, gestures, tables, graphs, and concrete objects. `GMP2.1`	19a, 20a
	Make sense of the representations you and others use. `GMP2.2`	19b, 20b–20f
SMP3	Make mathematical conjectures and arguments. `GMP3.1`	11b, 13e
	Make sense of others' mathematical thinking. `GMP3.2`	7a, 10g
SMP4	Model real-world situations using graphs, drawings, tables, symbols, numbers, diagrams, and other representations. `GMP4.1`	1–4, 19a, 20a
	Use mathematical models to solve problems and answer questions. `GMP4.2`	1–4, 19b, 19c, 20b–20f
SMP6	Explain your mathematical thinking clearly and precisely. `GMP6.1`	9e, 10g, 11b, 14c, 24b

End-of-Year Assessment

For Problems 1–4, write a number model. Use a ? to show what you
need to find out. Solve the problem. Write the answer.

① The second-grade class has 25 children. The first-grade class
has 20 children. How many children are there in all?

Number model: _____

There are _____ children.

② There are 42 cars in the parking lot. 30 cars leave. How many
cars are left in the parking lot?

Number model: _____

There are _____ cars left.

③ Shawn has 24 crayons. His teacher gave him 24 more. Then he
lost 8 crayons. How many crayons does he have now?

Number model(s): _____

Shawn has _____ crayons.

End-of-Year Assessment (continued)

④ 45 children are on the bus. At the first stop, 8 more children get on the bus. At the next stop, 16 children get off the bus. How many children are on the bus now?

Number model(s): _____

There are _____ children.

⑤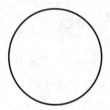

a. Divide the circle into three equal parts.

b. Use words to name one part.

_____

c. Use words to name all the parts.

_____

⑥ a. Write any 3-digit number: _____ _____ _____

b. Write the value of each digit in your number.

The value of _____ is _____.

The value of _____ is _____.

The value of _____ is _____.

End-of-Year Assessment (continued)

(7) Keon and Dartrianna showed the same number with base-10 blocks.

Keon's way   Dartrianna's way

a. What is the number? _____

b. Use base-10 shorthand to show this number another way.

c. Whose way makes it easiest to tell what the number is? Explain. _____

_____

(8) I have a 0 in my tens place.

I have a 7 in my hundreds place.

I have a 0 in my ones place.

What number am I? _____

(9) Write <, >, or =.

a. 89 _____ 88   b. 421 _____ 419

c. 709 _____ 790   d. 934 _____ 943

e. Explain how you got your answer to Problem 9d.

_____

_____

End-of-Year Assessment (continued)

(10) Solve.

a.   56
   + 37

b.   342
   + 139

c.   64
   − 39

d.   256
   − 178

e. 13 + 27 + 20 =

f. 31 + 22 + 19 =

g. Below, Marsha explained how she solved Problem 10d.

*I counted up from 178 to 200. I knew that was 22.
I know that from 200 to 256 is 56 so I added 22 and
56 and got 78.*

Does Marsha's strategy work? _____ Explain.

_____

_____

_____

_____

End-of-Year Assessment (continued)

11 **a.** Circle the largest number.

3,241   3,421   3,204   3,021

b. Parker wrote the numbers in order from smallest to largest like this:

3,204   3,021   3,241   3,421

smallest   largest

Do you agree with how Parker ordered the numbers? _____

Explain. _____

_____

_____

c. How would you write the numbers from smallest to largest?

_____   _____   _____   _____

smallest   largest

12 Complete the table.

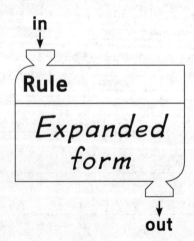

in	out
78	70 + 8
33	30 + 3
260	
496	

End-of-Year Assessment (continued)

13 Estimate the length of the line segment below to the nearest inch and to the nearest centimeter.

_____

a. Estimate: about _____ inches

b. Estimate: about _____ centimeters

Measure the length of the line segment to the nearest inch and centimeter.

c. about _____ inches   **d.** about _____ centimeters

e. Does the measurement have more inches or more centimeters? Explain your answer. _____

_____

_____

_____

14 Line Segment A: ——————
Line Segment B: ———————————————

a. Measure line segment A above to the nearest centimeter.
about _____ cm

b. How much longer is line segment B than the line segment A?
about _____ cm

c. How did you find your answer for part b? _____

_____

_____

End-of-Year Assessment (continued)

(15) A blue ribbon is 20 centimeters long. A yellow ribbon is 38 centimeters long.

How much longer is the yellow ribbon than the blue ribbon?

Number model: _____

The yellow ribbon is _____ centimeters longer than the blue ribbon.

(16) Solve the problem. Show your thinking on an open number line.

Christy has 43 green blocks. Ella has 36 yellow blocks. How many blocks do they have in all? _____ blocks

End-of-Year Assessment (continued)

(17) Write the time.

a.

b.

_____ : _____

_____ : _____

(18) Use ⓠ, ⓓ, ⓝ, **and** ⓟ to show two ways to pay for the bag of pretzels.

Pretzels
16 oz
99¢

One Way

Another Way

End-of-Year Assessment (continued)

(19)   Length of Pencils in Room 325

Length	Number of Pencils
2 inches	/
4 inches	//
6 inches	~~////~~ /
7 inches	///

a. Show the pencil lengths on the line plot.

Number
of
Pencils

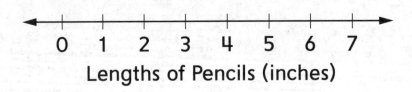

Lengths of Pencils (inches)

b. Write a question that can be answered using the line plot.

_____

_____

c. Write the answer to your question. _____

End-of-Year Assessment (continued)

Number of Books Read

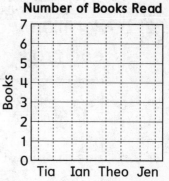

20 **a.** Complete the graph.

Tia read 6 books.
Ian read 3 books.
Theo read 5 books.
Jen read 3 books.

b. Who read the most books? _____

c. Who read the fewest books? _____

d. Who read more books—Ian or Theo? _____

e. How many more? _____

f. How many books did the children read altogether?

_____ books

21 **a.** Write at least one name for the shape.

_____

b. Draw a different shape with 4 sides and 2 pairs of parallel sides.

End-of-Year Assessment (continued)

22 **a.** Partition the rectangle into 4 rows with 6 same-size squares in each row.

    b. How many squares did you draw? _____

    c. How do you know your answer is correct?

    _____

    _____

23

    a. How many dots? _____

    b. Write an addition number model. _____

End-of-Year Assessment (continued)

24 **a.** Draw lines to divide the shape below into two equal parts.

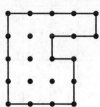

b. Explain how you know the two parts are equal.

_____

_____

_____

25 Circle the tool that you would use to measure the length of a bus.

     a six-inch ruler

     a yardstick

     a tape measure

     a meterstick

Explain why you chose that tool.

_____

_____

_____

_____

Teacher Facts Record Sheet

Column headers (facts):
10 + 4, 2 + 6, 10 + 3, 2 + 5, 10 + 6, 2 + 4, 10 + 7, 2 + 3, 10 + 8, 10 + 9, 4 + 6, 3 + 7, 2 + 8, 1 + 9, 0 + 10, 10 + 10, 9 + 9, 8 + 8, 7 + 7, 6 + 6, 5 + 5, 4 + 4, 3 + 3, 2 + 2, 1 + 1

Names

1.
2.
3.
4.
5.
6.
7.
8.
9.
10.
11.
12.
13.
14.
15.
16.
17.
18.
19.
20.
21.

Teacher Facts Record Sheet: (continued)

Names	3 + 6	3 + 4	7 + 8	6 + 9	4 + 8	6 + 7	5 + 6	4 + 7	6 + 8	3 + 8	4 + 5	5 + 8	7 + 9	3 + 6	5 + 7	8 + 9	4 + 9	5 + 6	3 + 5	2 + 10	2 + 9	10 + 5	2 + 7	
1.																								
2.																								
3.																								
4.																								
5.																								
6.																								
7.																								
8.																								
9.																								
10.																								
11.																								
12.																								
13.																								
14.																								
15.																								
16.																								
17.																								
18.																								
19.																								
20.																								
21.																								

Unit ____ Assessment Check-Ins
Individual Profile of Progress

Lesson	Standards	Assess Progress	Comments

Assess Progress   **M** = meeting expectations   **N** = not meeting expectations   **N/A** = not assessed

Unit ___ Assessment Check-Ins: Class Checklist

	Assessment Check-In Items by Lesson										
Names											

Assess Progress   **M** = meeting expectations   **N** = not meeting expectations   **N/A** = not assessed

Unit ___ **Progress Check**
Individual Profile of Progress

Item(s)	Standards	Assess Progress	Comments
Unit Assessment			
Open Response Assessment (odd-numbered units)			

Assess Progress   **M** = meeting expectations   **N** = not meeting expectations   **N/A** = not assessed

NAME

DATE

Unit ___ Progress Check
Individual Profile of Progress

Item(s)	Standards	Assess Progress	Comments
Challenge (Optional)			

Assess Progress   **M** = meeting expectations   **N** = not meeting expectations   **N/A** = not assessed

✂ –

NAME

DATE

Unit ___ Progress Check
Individual Profile of Progress

Item(s)	Standards	Assess Progress	Comments
Challenge (Optional)			

Assess Progress   **M** = meeting expectations   **N** = not meeting expectations   **N/A** = not assessed

Unit ____ **Progress Check:** Class Checklist

Unit Assessment Items									Challenge (Optional)	Open Response Assessment
Names										

Assess Progress   **M** = meeting expectations   **N** = not meeting expectations   **N/A** = not assessed

Unit ____ Progress Check
Individual Profile of Progress

Item(s)	Standards	Assess Progress	Comments
Cumulative Assessment			

Assess Progress   **M** = meeting expectations   **N** = not meeting expectations   **N/A** = not assessed

Unit ____ Progress Check: Class Checklist

	Cumulative Assessment Items										
Names											

Assess Progress   **M** = meeting expectations   **N** = not meeting expectations   **N/A** = not assessed

Mathematical Process and Practice for Unit(s) _____
Individual Profile of Progress

Use this sheet to record children's use of the mathematical processes and practices in lesson activities, Assessment Check-Ins, Writing/Reasoning Prompts, and Progress Checks.	Opportunity	Date	Comments
SMP1: Make sense of problems and persevere in solving them.			
GMP1.1 Make sense of your problem.			
GMP1.2 Reflect on your thinking as you solve your problem.			
GMP1.3 Keep trying when your problem is hard.			
GMP1.4 Check whether your answer makes sense.			
GMP1.5 Solve problems in more than one way.			
GMP1.6 Compare the strategies you and others use.			
SMP2: Reason abstractly and quantitatively.			
GMP2.1 Create mathematical representations using numbers, words, pictures, symbols, gestures, tables, graphs, and concrete objects.			
GMP2.2 Make sense of the representations you and others use.			
GMP2.3 Make connections between representations.			
SMP3: Construct viable arguments and critique the reasoning of others.			
GMP3.1 Make mathematical conjectures and arguments.			
GMP3.2 Make sense of others' mathematical thinking.			
SMP4: Model with mathematics			
GMP4.1 Model real-world situations using graphs, drawings, tables, symbols, numbers, diagrams, and other representations.			
GMP4.2 Use mathematical models to solve problems and answer questions.			

Mathematical Process and Practice
Individual Profile of Progress (continued)

	Opportunity	Date	Comments
SMP5: Use appropriate tools strategically.			
GMP5.1 Choose appropriate tools.			
GMP5.2 Use tools effectively and make sense of your results.			
SMP6: Attend to precision.			
GMP6.1 Explain your mathematical thinking clearly and precisely.			
GMP6.2 Use an appropriate level of precision for your problem.			
GMP6.3 Use clear labels, units, and mathematical language.			
GMP6.4 Think about accuracy and efficiency when you count, measure, and calculate.			
SMP7: Look for and make use of structure.			
GMP7.1 Look for mathematical structures such as categories, patterns, and properties.			
GMP7.2 Use structures to solve problems and answer questions.			
SMP8: Look for and express regularity in repeated reasoning.			
GMP8.1 Create and justify rules, shortcuts, and generalizations.			

Mathematical Process and Practice Opportunities
Class Record

Standard for Mathematical Process and Practice _____

Goal for Mathematical Process and Practice _____

Opportunity: _____

Names	+ / ✔ / –	Comments

Mathematical Process and Practice in Open Response Problems
Individual Profile of Progress

Lesson	SMP	GMP	Assess Progress	Comments
Open Response and Reengagement				
1-5	7	7.2		
2-7	8	8.1		
3-1	6	6.1		
4-6	2	2.2		
5-11	1	1.5		
6-9	5	5.2		
7-2	7	7.2		
8-4	3	3.1		
9-3	4	4.2		
9-9	6	6.1		
Progress Check				
1-13	2	2.1		
3-12	3	3.2		
5-12	3	3.1		
7-10	2	2.2		
9-12	3	3.1		

Assess Progress   **E** = exceeding expectations   **M** = meeting expectations

P = partially meeting expectations   **N** = not meeting expectations

Mathematical Process and Practice in Open Response Problems
Class Checklist

	Open Response and Reengagement Items										Progress Check Items				
Names	SMP7 GMP7.2 **1-5**	SMP8 GMP8.1 **2-7**	SMP6 GMP6.1 **3-1**	SMP2 GMP2.2 **4-6**	SMP1 GMP1.5 **5-11**	SMP5 GMP5.2 **6-9**	SMP7 GMP7.2 **7-2**	SMP3 GMP3.1 **8-4**	SMP4 GMP4.2 **9-3**	SMP6 GMP6.1 **9-9**	SMP2 GMP2.1 **1-13**	SMP3 GMP3.2 **3-12**	SMP3 GMP3.1 **5-12**	SMP2 GMP2.2 **7-10**	SMP3 GMP3.1 **9-12**

Assess Progress　　**E** = exceeding expectations　　**M** = meeting expectations

P = partially meeting expectations　　**N** = not meeting expectations

NAME

DATE

My Exit Slip

NAME

DATE

My Exit Slip

About My Math Class A

Draw a face or write the words that show how you feel.

 Good   OK   Not so good

① This is how I feel about math:	② This is how I feel about working with a partner or in a group:	③ This is how I feel about working by myself:
④ This is how I feel about solving number stories:	⑤ This is how I feel about doing Home Links with my family:	⑥ This is how I feel about finding new ways to solve problems:

Circle **yes, sometimes,** or **no.**

⑦ I like to figure things out. I am curious.

yes   sometimes   no

⑧ I keep trying even when I don't understand something right away.

yes   sometimes   no

About My Math Class B

Circle the word that best describes how you feel.

(1) I enjoy mathematics class.   **yes   sometimes   no**

(2) I like to work with a partner   **yes   sometimes   no**
or in a group.

(3) I like to work by myself.   **yes   sometimes   no**

(4) I like to solve problems   **yes   sometimes   no**
in mathematics.

(5) I enjoy doing Home Links   **yes   sometimes   no**
with my family.

(6) In mathematics, I am good at _____

_____

_____

_____

(7) One thing I like about mathematics is _____

_____

_____

(8) One thing I find difficult in mathematics is _____

_____

_____

Math Log A

What did you learn in mathematics this week?

_____

_____

_____

_____

_____

_____

 -

Math Log A

What did you learn in mathematics this week?

_____

_____

_____

_____

_____

_____

Math Log B

Question: _____

_____

_____

_____

_____

_____

_____

_____

Math Log B

Question: _____

_____

_____

_____

_____

_____

_____

NAME

DATE

Math Log C

Work Box	Tell how you solved this problem.
	_____
	_____
	_____
	_____
	_____
	_____

✂ -

NAME

DATE

Math Log C

Work Box	Tell how you solved this problem.
	_____
	_____
	_____
	_____
	_____
	_____

NAME	DATE	

Good Work!

 I have chosen this work because _____

_____

_____

_____

_____

_____

- -

NAME	DATE	

Good Work!

 I have chosen this work because _____

_____

_____

_____

_____

_____

NAME   DATE

My Work

This work shows that I can _____

_____

_____

I am still learning to _____

_____

_____

_____

_ _

NAME   DATE

My Work

This work shows that I can _____

_____

_____

I am still learning to _____

_____

_____

_____

Interim Assessment Answers

Beginning-of-Year Assessment

1. 30, 31, 33

2. 65, 75, 85, 90

3. 75, 85, 115

4. 9

5. 70

6.

6	7	8	9
	17	18	
		28	29

7.

40	41	42	43
	51	52	53
60		62	63

8. 50, 60, 80, 100

9. a. 8 **b.** 5 **c.** 10 **d.** 12 **e.** 2 **f.** 5

10. a. = **b.** < **c.** >

11. Sample answers: $5 + 5$, $6 + 4$, $7 + 3$, $8 + 2$, $9 + 1$, $10 + 0$, ten, $12 - 2$, $15 - 5$, $20 - 10$

12. 7; $4 + 3 = 7$ or $4 + 3 = $ _____

13. 8; $16 - 8 = 8$ or $16 - 8 = $ _____ or $8 + 8 = 16$ or $8 + $ _____ $= 16$

14. a. 3:00 **b.** 1:30

Mid-Year Assessment

1. a. 6 **b.** 1 **c.** 3 **d.** 1

2. a. 14 **b.** 11 **c.** 10 **d.** 16
e. $2 + 9$;
Sample answer: It is not a doubles fact.

3. $8 + 7 = 15$, $7 + 8 = 15$, $15 - 7 = 8$, $15 - 8 = 7$

4. Order varies. $0 + 10$, $10 + 0$, $1 + 9$, $9 + 1$, $2 + 8$, $8 + 2$, $3 + 7$, $7 + 3$, $4 + 6$, $6 + 4$, $5 + 5$

5. a. even
b. Sample answer: I matched each star with another star and there were no stars left over.
c. $7 + 7 = 14$

6. a.

96 _____ 98 _____ 100 _____ 102 _____ 104 _____ 106

b. Sample answers: They are all odd numbers. They are counting up by 2s.

7.

0 — 10 — 25

8. a.

47		

b.

		99

c. Sample answer: The digit in the ones place stays the same. The digit in the tens place goes up by one each time.

9. a. 534; 543
b. Sample answer: 537;
Sample answer: five hundred thirty-seven

10. +10; 100, 130, 220

11. 35¢;
Sample answer: Ⓓ Ⓓ Ⓓ Ⓝ
Sample answer: Ⓝ Ⓝ Ⓝ Ⓓ Ⓓ

12. Sample answer: Ⓓ Ⓓ Ⓓ Ⓝ Ⓝ Ⓟ Ⓟ Ⓟ Ⓟ

13. Yasmin; Sample answer: The clock says 3:45 because the hour hand is between the 3 and the 4 and the minute hand is on the 9. Sample answer: He mixed up the hour and minute hands.

14. a. afternoon
b.

End-of-Year Assessment

1. $25 + 20 = ?; 45$

2. Sample answers: $42 - 30 = ?, 42 = 30 + ?; 12$

3. Sample answers: $24 + 24 = 48$ and $48 - 8 = ?,$
$24 + 24 - 8 = ?; 40$

4. Sample answers: $45 + 8 = 53$ and $53 - 16 = ?,$
$45 + 8 - 16 = ?; 37$

5. a.

   b. Sample answer: one-third

   c. Sample answer: three-thirds

6. a. Answers vary.   **b.** Answers vary.

7. a. 234

   b. Sample answer:

   c. Sample answer: Keon's way because it uses
fewer blocks.

8. 700

9. a. $>$ **b.** $>$ **c.** $<$ **d.** $<$

   e. Sample answer: 934 has one less 10 than 943.

10. a. 93   **b.** 481   **c.** 25   **d.** 78   **e.** 60   **f.** 72

   g. Yes. Sample answer: She added numbers
to get to easier numbers. 178 plus 2 is 180,
plus 20 more is 200, plus 56 is 256.
$56 + 2 = 58$, 58 plus 20 is 78.

11. a. 3,421

   b. No. Sample answer: 3,021 is the smallest
number because it doesn't have any
hundreds, but 3,204 has 2 hundreds.

   c. 3,021; 3,204; 3,241; 3,421

12. $200 + 60 + 0; 400 + 90 + 6$

13. a. Answers vary.   **b.** Answers vary.

   c. 4   **d.** 11

   e. There are more centimeters in the
measurement because centimeters are
smaller units than inches.

14. a. 2   **b.** 5

   c. Sample answer: I measured the difference
from the end of Line Segment A to the end
of Line Segment B.

15. Sample answers: $20 + ? = 38, 38 - 20 = ?; 18$

16. 79; Sample answer:

17. a. 8:05   **b.** 4:25   **18.** Answers vary.

19. a.

Number
of
Pencils

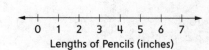

Lengths of Pencils (inches)

   b. Sample answer: How many pencils are
7 inches long?

   c. Sample answer: 3 pencils are 7 inches long.

20. a.

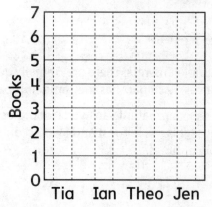

   b. Tia   **c.** Ian and Jen

   d. Theo   **e.** 2   **f.** 17

End-of-Year Assessment (continued)

21. a. Sample answers: quadrilateral, square

   b. Sample answer:

22. a.

   b. 24

   c. Sample answer: I counted to make sure there are 4 rows with 6 squares in each.

23. a. 16

   b. $8 + 8 = 16$ or

      $2 + 2 + 2 + 2 + 2 + 2 + 2 + 2 = 16$

24. a.

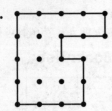

   b. Sample answers: There are 6 small squares in each of the halves. Each part has the same number of little squares.

25. Answers vary. Sample answer: I would use the tape measure because it is the longest and I would only need to move it a couple of times.